SPIRIT OF
SUFFOLK
WINDMILLS

CHRIS HERRING

First published in Great Britain in 2011

British Library Cataloguing-in-Publication Data
A CIP record for this title is available from the British Library

ISBN 978 0 85710 025 2

PiXZ Books
Halsgrove House, Ryelands Business Park,
Bagley Road, Wellington, Somerset TA21 9PZ
Tel: 01823 653777
Fax: 01823 216796
email: sales@halsgrove.com

An imprint of Halstar Ltd, part of the Halsgrove group
of companies Information on all Halsgrove titles is
available at: www.halsgrove.com

Printed and bound in China by Toppan Leefung Printing Ltd

Introduction

When it comes to windmills and drainage mills Suffolk is often overlooked in deference to neighbouring Norfolk; however Suffolk too has a wonderful selection of iconic windmills and drainage mills, as this book reveals.

In the early nineteenth century almost five hundred mills were at work in the Suffolk countryside, but due to more efficient steam and diesel engines being introduced many of the windmills were abandoned and slowly fell into disrepair. Today there are only a handful of Suffolk mills that are in working condition and still producing flour.

Despite the decline in mills and milling from the late nineteenth century onwards, Suffolk can still boast a remarkable variety of survivors, some still retaining their original gears and millstones. Now part of our heritage, a number of these mills can be visited by the general public. From the picturesque drainage mill on the Herringfleet Marshes to the superb working post mill at Stanton, these wonderful buildings punctuate the landscape for miles around and add to the beauty of our unique Suffolk countryside.

Bardwell Windmill is believed to have been built sometime in the 1820s. The mill stands at the edge of the village and served the local community until the 1940s when it fell into dereliction. With the help of local volunteers, restoration work has seen the mill fitted with a new cap and tail fan, with work beginning on a set of new sails.

Built in 1836 Buttram's Mill is one of England's finest tower windmills. The mill, located in Woodbridge, was built by the famous Suffolk millwright John Whitmore.

Left:
Framsden Post Mill, built in 1760, stands high on a hill overlooking the small village of Framsden. The original appearance of the mill would have been very different to how it looks now. In 1836 the mill was raised eighteen feet when the brick round-house at the base of the mill was added. It is thought the mill finally ceased working in 1936 and gradually fell into a state of dereliction before a group of volunteer's began restoring it in 1966.

Opposite:
The first floor of Framsden Post Mill decorated with a selection of flour sacks.

Above left: Grain sacks on the sack floor at Framsden Post Mill
Above right: A pair of millstones inside Framsden Post Mill.

Opposite:
The vertical post in Framsden Post Mill, to the right hangs a metal chain
which was used to hoist the sacks of grain up the mill through various trapdoors.

One of Framsden Post Mill's grain hoppers shown here in restored condition.

The brake wheel and sack hoist drive seen here inside Framsden Post Mill.

Above: Here a local volunteer can be seen painting the mill at Friston.

Opposite: Friston Post Mill is one of the tallest post mills in England. What makes this mill unique is that it was originally erected on Mill Hill in Woodbridge before being moved to its present site in the village of Friston in 1812. In 1965 permission was granted for the demolition of the mill, although thankfully this was never carried out and over the years work has gradually taken place to help preserve it.

Opposite:
Collis Mill is a three-storey smock mill
sited at Great Thurlow. It is thought
the mill was moved to its current site
in 1807 and it continued to work until the
1920s, after which it gradually fell into a
derelict state before restoration work began
in 1959. Here the mill can be seen surrounded
by scaffolding as further maintenance
and preservation work is carried out.

Right:
Herringfleet Drainage Mill
illuminated by the evening sun.

Overleaf:
Of all the mills in Suffolk, Herringfleet is
certainly among the most picturesque.
The mill itself stands in an isolated spot on
marshland beside the River Waveney. Beneath
wide skies these flatlands are criss-crossed
by lines of open drainage ditches.

Left:
Here the mill at Herringfleet can be
seen at dawn following a winter hoarfrost.

Opposite:
The wooden smock drainage
mill at Herringfleet was erected c.1920
by Great Yarmouth-based millwright
Robert Barnes. The mill worked by wind-
power until 1956. It was transferred to the
ownership of the County Council in 1958.
In this photograph the mill is captured
at sunrise on a wonderful winter's morning.

Here the mill at Herringfleet is seen at sunrise on a cold February morning. What I like most about this image is the backlit reeds showing off the wonderful detail of the tiny ice crystals brought on by winter hoarfrost.

The marshland around Herringfleet Mill is transformed into a winter wonderland as a result of freezing mist and fog depositing delicate ice crystals on the surrounding landscape.

Opposite: Herringfleet Mill was built to drain the surrounding marshland. Such drainage mills are common in Suffolk and differ from the type of mills constructed to grind flour.

Left:
Herringfleet Mill reflecting in a nearby
dyke captured here at last light on
a Spring afternoon.

Opposite:
Herringfleet Mill lies inside the Norfolk
and Suffolk Broads National Park.
The mill itself lies twenty yards on the
Suffolk side of the county's border
with Norfolk.

The silhouette of Herringleet Mill at sunset.

Opposite:
This view of Herringfleet Mill shows off the wonderful colours of the rusty old storage shed located next to the mill. Like most drainage mills on the Norfolk and Suffolk Broads when, in the first half of the twentieth century, it became more efficient to drain the land with more modern diesel engines, many of the mills were converted to run off on diesel power. It is possible the corrugated iron shed would have housed such equipment.

Right:
The mill at Herringfleet is winded by a tail pole chain winch; this required the marshman to manually manoeuvre the sails to face the oncoming wind. Here the tail pole can be seen at the back of the mill.

Opposite:
Holton Windmill sits on an elevated location overlooking the village of Holton. The mill was originally built in 1749 by John Swann. It is thought it last worked around 1910, shortly after which most of the internal workings were removed and the mill was used as a summer house. In 1949 restoration work began and, although a great deal of care has gone into the restoration of the exterior of the mill, only the windshaft and brake wheel remain of the internal machinery.

Right:
During the restoration of the mill a local millwright inspected the mill for the Ministry of Works. As the mill was not structurally sound enough to continue working it was decided to restore it as a landmark. The mill itself once carried two spring sails and two common sails, the current sails were added to the mill in 1991.

A millstone lies against the brickwork of the roundhouse at Holton Windmill.

Although when originally built the mill at Holton would not have had a fantail one was added at a later date. Unfortunately when I visited the mill the fantail had been removed for repairs; it would normally sit between the two posts in the top right corner, automatically turning the sails to face the oncoming wind.

Pakenham remains a working mill and continues to produce flour. Electricity is also used as a power source and here some of the internal equipment can be seen.

Opposite:
Pakenham Windmill is a five storey brick-built tower mill constructed in 1831. The mill itself is eighty feet tall and has a dome-shaped cap and tailfan.

Above left: The brake wheel and windshaft of Pakenham Windmill.
Above right: Milling equipment inside Pakenham Windmill.

Opposite:
Pakenham Windmill. In the middle of the photograph one of the grain hoppers can be seen feeding grain on to a pair of millstones.

Weighing scales on the first floor at Pakenham Windmill.

A flour bin seen here on the first floor at Pakenham Mill.

Saxtead Green is a post mill that dates back to at least 1796 when the miller was then Amos Webber. The mill contains four patent sails and is winded by a fantail. During its lifetime the height of the mill has been raised three times. Here the mill is viewed from the nearby road that passes through the village.

Opposite:
Skoulding's Mill is a seven-storey tower mill located in the village of Kelsale. The mill was originally built in 1856 and was last thought to have worked by wind power in the early twentieth century. In the 1950s the cap and much of the internal workings were removed before being converted into residential accommodation.

Right:
This fantastic small brick-built mill, known as Blackshore Mill, can be found at Reydon on the Southwold side of the River Blyth. It is photographed here on a bright spring afternoon.

The interior of Blackshore Mill, lying close to the River Blyth, is in a pretty poor state; here the timbers of the collapsed first floor can be seen. The mill was originally built c.1890 by the Beccles millwright Robert Martin. The mill had a very short life, becoming damaged after just a few years of use; the mill remained in a derelict state until 2002 when some repairs were carried out to help conserve it.

Blackshore Mill on a sunny June afternoon. A five-minute
walk from this mill brings you to picturesque Southwold Harbour.

The mill at Stanton is still regularly producing flour and is one of only a handful of post mills still working in the country today.

Opposite: Originally built in 1751 on the opposite side of Stanton village, the mill was moved to its current location around 1818. The fantail and brick roundhouse would have been added at a later date. The mill is currently owned by Dominic and Linda Grixti who work tirelessly to keep it in a fantastic working condition.

Above left: Here parts of the sails of Stanton Post Mill have been removed for painting and are drying in the sun.
Above right: The millstones in Stanton Post Mill.

Opposite: Here the brake wheel can be seen in the background of the photograph, with the sack hoist in the foreground. The hoist is used to haul up bags of grain through various trapdoors all the way to the top of the mill.

Sadly Syleham Mill is in a derelict state with only the roundhouse remaining. It was originally built in 1730 next to another post mill at Wingfield Green. The mill was moved to its present location c.1823. Today the roundhouse is protected by being clad in corrugated iron. What is particularly fascinating about this mill is that it is still possible to see many of the original parts such as sails and internal workings scatted around the outside of the mill.

Here the mill at Thelnetham is seen against a dark stormy sky
on a spring morning in the Suffolk countryside.

Opposite:
Thelnetham Tower Mill was built in the early nineteenth century to grind wheat into
flour. It is thought the mill worked until the 1920s after which it fell into a derelict
state. In 1979 a group of enthusiasts purchased the mill and began restoration.

Opposite:
Thorpness is a grade II listed post mill located in the popular holiday resort of Thorpness. The mill was originally built in 1803 in the neighbouring village of Aldringham to grind corn into flour. In 1923 the mill was moved to Thorpness where it began a new lease of life as windpump to pump water up to the nearby water tower. Today this tower is used as holiday accommodation and is known as the 'House in the Clouds'.

Right:
The mill at Thorpness continued to work as a water pump until 1940, after which it was replaced by an engine. This too was made redundant in 1963 when the village was joined to the water mains.

Today the mill at Thorpness is open to the public who can explore its fascinating workings.

Opposite:
The sails of Thorpness Mill seen against a cloudy summer sky.

Original gears and mill equipment can be seen on display inside the mill at Thorpness.

Opposite:
The brake wheel and windshaft inside Thorpness Mill.

Previous page:
Westwood Marshes Mill sits amid empty marshland close to the coastal resort of Walberswick. The mill was built in the late eighteenth century and continued to work by wind until 1940. It was damaged after being used as a coastal gunnery target during the war. The mill was repaired in the 1950s but then became the victim of an arson attack in the 1960s.

Left:
Today the mill at Walberswick stands in a derelict state towering above the reeds and marshes, making it an useful navigation point for walkers and birdwatchers.